LIVY LITTLE
LEARN TO COUNT WITH LIVY LITTLE

Gabriella Gugliotta-Comes

Copyright © 2015 by Gabriella Gugliotta-Comes

All rights reserved. This book or any portion thereof may not be reproduced or used in any manner whatsoever without the express written permission of the publisher except for the use of brief quotations in a book review.
Printed in the United States of America
First Printing, 2015

Library of Congress Control Number: 2015954054
ISBN 978-0-9969108-0-4

ABC Book Publishers, Inc.
4940 Blackhawk Drive
St. Johns, FL 32259
abcbookpublishers.com

This is for all the children who see the beauty in "What wonders can be done with numbers".

To my Michael and my LiviaRose, (my Mona Lisa) —We Live, We Laugh, My Loves!

To one of my best friends, Deana Gioia, Thank you for always supporting me!---G.C.

Livy Little is a little bear which always has a twitch with an itch to enrich in numbers. She does not like to miscount because then she has to recount, but never-the-less, this is what keeps Livy Little in shipshape.

She thinks to herself, hummm… numbers are just a real railroad of delight! "What wonders that can be done with numbers!". Numbers are all around. There are endless and boundless ways you can count with numbers!

I can count my toes and fingers, wiggling away. 1, 2, 3 and 4, wow, my eyelashes and sashes and more, again 1, 2, 3, and 4

Wow... "What wonders can be done with numbers!"

I can count my hats and doormats. I can also count the 1, 2 and 3 mice that chat with the three cats that seem to scat when the three mice pat on the doormat.

Enjoying the morning, Livy Little counts the days, weeks, and maybe the months, by constantly saying out loud—"What wonders that can be done with numbers."

She brushed her teeth and hair at the beginning of the day. 1 stroke, 2 strokes, and 3 strokes more.

Wow, my hair and teeth are shiny as can be. I have done two chores, during the beginning of this fine bright day. She thought to herself, "What wonders can be done with numbers!"

Rumble, Rumble went her stomach as Livy Little looked down. Breakfast is what came to mind, "Hummm....what shall I have?" she said out loud. Maybe some cereal and apples, or eggs with cheese, something that will keep me appeased. Maybe some peanuts and cheese is what I might need? What munchies that I have when wondering with numbers.

Livy Little put the cereal with apples and the eggs with cheese. She counted profusely **1** + **2**...**3**, **4** and **5**, but if I count the peanuts and cheese, add one more, it becomes 10 for sure.

Ahhh this is not correct, what have I done is a glitch with a ripple that caused a flutter, which is not 10 for sure.. **1**, **2**, **3**, **4** and **5** she counted her fingers and toes... hummm...I found where I counted incorrectly!

I miscounted and noted, by putting a **1** there and another number here is how I arrived at the answer which put me in a pickle!

The correct number I applied and tried to tie with cereal, apples, eggs with cheese and yes, certainly, I can put peanuts plus cheese. Now then, the number I seek is **7**, what a fix. "I am correct!" Livy Little said out loud.

When miscounting takes place, I see all the damage that occurs.

I can't go all week eating blueberry pie nor will I have to fry? Who knows? This is why I must keep my eye on the prize, and not rely on anything else except my mind. It is the knowledge that gives me the passage and privilege to count. "What wonders I have done with numbers!"

In the kitchen is where I whip up my best fries, and 1, 2, 3, and 4 pies. I can even count the butterflies.

What does all of this total? It equals 4 pies, 3 fries and 4 butterflies. I must share with my friends this wondrous superb saying, "What wonders can be done with numbers!"

Thinking away, I have done so much today, here are just a few of the fun things— Brushed my teeth and hair; it's my routine. Sparkly and clean is how I love to be.

I also, counted my hats, doormats on which I noticed the 3 mice who chat with the 3 cats. Baking pies, I caught some butterflies. Go and spend your day counting in multiple ways and read about all the possibilities that will keep you at play. Adding numbers and how they are used is how Livy Little is amused.

s I call upon my friends one by one, I heard a new sound that was unbound. I looked out my window and to my surprise was a spotted bird before my very own eyes.

"Join us on the tale of the spotted bird. We'll take you on a counting journey you will never forget! We'll count 1, 2, 3, and 4 clouds that pass us by, in the meanwhile, let's have fun as we say goodbye!"

"What wonders can be done with numbers!"

Made in the USA
Middletown, DE
04 November 2022

14048604R00015